D1037516

THE READING LIFE

ALSO BY C. S. LEWIS

A Grief Observed

George MacDonald: An Anthology

Mere Christianity

Miracles

The Abolition of Man

The Great Divorce

The Problem of Pain

The Screwtape Letters
(with "Screwtape Proposes a Toast")

The Weight of Glory

The Four Loves

Till We Have Faces

Surprised by Joy: The Shape of My Early Life

Reflections on the Psalms

Letters to Malcolm: Chiefly on Prayer

Personal Heresy

The World's Last Night & Other Essays

Poems

The Dark Tower & Other Stories

Of Other Worlds: Essays & Stories

Narrative Poems

Letters of C. S. Lewis

All My Road Before Me

The Business of Heaven: Daily Readings from C. S. Lewis

Present Concerns: Essays by C. S. Lewis

Spirits in Bondage: A Cycle of Lyrics

On Stories: And Other Essays of Literature

ALSO AVAILABLE FROM HARPERCOLLINS

The Chronicles of Narnia

The Magician's Nephew

The Lion, the Witch and the Wardrobe

The Horse and His Boy

Prince Caspian

The Voyage of the Dawn Treader

The Silver Chair

The Last Battle

C.S. Lewis

THE READING LIFE

The Joy of Seeing New Worlds Through Others' Eyes

EDITED BY DAVID C. DOWNING AND MICHAEL G. MAUDLIN

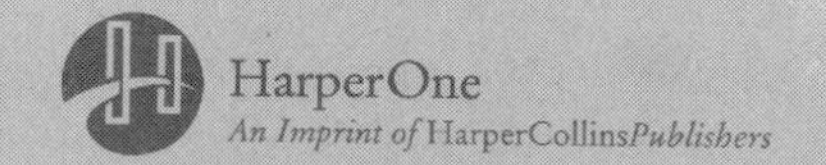

HarperOne

HarperCollins books may be purchased for educational, business, or sales promotional use. For information, please email the Special Markets Department at SPsales@harpercollins.com.

FIRST EDITION

Designed by Janet Evans-Scanlon

Library of Congress Cataloging-in-Publication Data has been applied for.

ISBN 978-0-06-284997-7

20 21 22 23 LSC 10 9 8 7 6 5 4 3 2

CONTENTS

PART ONE: ON THE ART AND JOY OF READING

PART TWO: SHORT READINGS ON READING

PREFACE

THE NOTED CRITIC WILLIAM EMPSON ONCE DESCRIBED C. S. Lewis as "the best-read man of his generation, one who read everything and remembered everything he read."[1] This sounds like pardonable exaggeration, but it comes close to being true in the realms of literature, philosophy, and classics. At the age of ten, Lewis started reading Milton's *Paradise Lost*. By age eleven, he began his lifelong habit of seasoning his letters with quotations from the Bible and Shakespeare. In his midteens, Lewis was reading classic and contemporary works in Greek, Latin, French, German, and Italian.

---

1 *C. S. Lewis at the Breakfast Table*, ed. by James Como (1992), xxiii.

And Lewis did indeed seem to remember most of what he read. One of his students recalled that someone could quote any line from the book-length *Paradise Lost*, and Lewis would continue the passage from memory. Another student said that he could take a book off Lewis's shelf, open a page at random and begin reading, and Lewis could summarize the rest of the page, often word for word.[2] With that kind of memory, Lewis had little difficulty reaching for just the right quotation or reference to illustrate his point. Since it seems he was able to carry an entire library in his head, it should come as no surprise that his major scholarly books average about one thousand citations apiece. His three volumes of letters contain another twelve thousand quotations or references. Even The Chronicles of Narnia for children contain nearly

---

2 Derek Brewer in Como, 47; Kenneth Tynan in Stephen Scho-field, *In Search of C. S. Lewis*, 6–7.

one hundred echoes or allusions to myth, history, or literature.

But as Mortimer Adler once remarked, "In the case of good books, the point is not to see how many of them you can get through, but rather how many can get through to you." Lewis would certainly agree, and he often commented how much his worldview and sensibility were shaped by the books he read—everything from Beatrix Potter in childhood to his re-reading of Homer's *Iliad*, Dickens's *Bleak House*, and Tennyson's *In Memoriam* in the last few weeks before his death in November 1963.

Lewis was a disciplined reader and an engaged reader. Fellow scholars recall how he could sit for hours in the Bodleian Library at Oxford, perusing and absorbing texts, oblivious to what was happening in the room around him. When reading books from his private library, he often added marginal notes and created his own index on the inside cover. If he found a book unprofitable, as he did Byron's *Don Juan*, he simply wrote on the inside back cover "Never again."

Of course, reading was also one of the supreme pleasures of Lewis's life. In his memoir *Surprised by Joy* Lewis described his ideal daily routine to be reading and writing from nine until one and again from five to seven, with breaks for meals, walking, or tea-time. Apart from those six hours of study every day, he also enjoyed light reading over meals or in the evening hours (pp. 141–143). All in all, Lewis's preferred schedule seemed to include seven or eight hours of reading per day! For Lewis, reading was both a high calling and an endless source of satisfaction. In fact, his sense of vocation and avocation were virtually indistinguishable whenever he picked up a book—and often when he wrote one.

Often Lewis described the community that is formed when one is among fellow passionate readers (see the chapter on "How to Know If You Are a

True Reader"). This fellowship is not one of merely sharing a hobby but of people whose worlds have been enlarged and deepened by books. They are a distinctive group. This collection brings together fun, whimsical, and wise selections from Lewis's lifetime of writing that would be of interest to those who share this passion. And we mean all who love reading literature, whether children's fantasy, poetry, science fiction, or Jane Austen. We did not include his opinions on classic or historical literature, which was his academic specialty, but only his advice and opinions on the shared enterprise of reading works of general interest. Nor do we include his many comments on Christian or devotional reading. This book is for members of the reading club, broadly defined.

One of the delights of Lewis's thoughts on reading is the breadth of his passions, never forgetting the childhood joy in discovering that books were portals to other worlds. As Lewis himself explained, "Literary experience heals the wound, without undermining the privilege, of individuality. . . . In reading great literature I become a thousand men and yet remain myself. Like the night sky in the Greek poem, I see with a myriad eyes, but it is still I who see. Here, as in worship, in love, in moral action, and in knowing, I transcend myself; and am never more myself than when I do."

This volume is for the entertainment and the edification of those in this reading club. We hope you enjoy this new window into the wit and wisdom of C. S. Lewis.

DAVID C. DOWNING
Codirector of the Marion E. Wade Center
at Wheaton College in Illinois

MICHAEL G. MAUDLIN
Senior Vice President and Executive Editor,
HarperOne

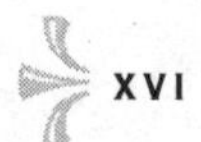

Part One

ON THE ART AND JOY OF READING

WHY WE READ

We seek an enlargement of our being. We want to be more than ourselves.

Each of us by nature sees the whole world from one point of view with a perspective and a selectiveness peculiar to himself. And even when we build disinterested fantasies, they are saturated with, and limited by, our own psychology. To acquiesce in this particularity on the sensuous level—in other words, not to discount perspective—would be lunacy. We should then believe that the railway line really grew

---

An Experiment in Criticism
(from the Epilogue)

narrower as it receded into the distance. But we want to escape the illusions of perspective on higher levels too.

We want to see with other eyes, to imagine with other imaginations, to feel with other hearts, as well as with our own. We are not content to be Leibnitzian monads. We demand windows. Literature as Logos is a series of windows, even of doors. One of the things we feel after reading a great work is 'I have got out'. Or from another point of view, 'I have got in'; pierced the shell of some other monad and discovered what it is like inside.

Good reading, therefore, though it is not essentially an affectional or moral or intellectual activity, has something in common with all three. In love we escape from our self into one other. In the moral sphere, every act of justice or charity involves putting ourselves in the other person's place and thus transcending our own competitive particularity. In

coming to understand anything we are rejecting the facts as they are for us in favour of the facts as they are. The primary impulse of each is to maintain and

We want to see with other eyes, to imagine with other imaginations, to feel with other hearts, as well as with our own.

aggrandise himself. The secondary impulse is to go out of the self, to correct its provincialism and heal its loneliness. In love, in virtue, in the pursuit of knowledge, and in the reception of the arts, we are doing this. Obviously this process can be described either as an enlargement or as a temporary annihilation of the self. But that is an old paradox; 'he that loseth his life shall save it'.

We therefore delight to enter into other men's beliefs (those, say, of Lucretius or Lawrence) even though we think them untrue. And into their passions, though we think them depraved, like those, sometimes, of Marlowe or Carlyle. And also into their imaginations, though they lack all realism of content.

This must not be understood as if I were making the literature of power once more into a department within the literature of knowledge—a department which existed to gratify our rational curiosity about other people's psychology. It is not a question of knowing (in that sense) at all. It is connaitre not savoir; it is erleben; we become these other selves. Not only nor chiefly in order to see what they are like but in order to see what they see, to occupy, for

a while, their seat in the great theatre, to use their spectacles and be made free of whatever insights, joys, terrors, wonders or merriment those spectacles reveal. Hence it is irrelevant whether the mood expressed in a poem was truly and historically the poet's own or one that he also had imagined. What matters is his power to make us live it. I doubt whether Donne the man gave more than playful and dramatic harbourage to the mood expressed in 'The Apparition.' I doubt still more whether the real Pope, save while he wrote it, or even then more than dramatically, felt what he expresses in the passage beginning 'Yes, I am proud'.

What does it matter?

This, so far as I can see, is the specific value or good of literature considered as Logos; it admits us to experiences other than our own. They are not, any more than our personal experiences, all equally worth having. Some, as we say, 'interest' us more than others. The causes of this interest are naturally extremely various and differ from one man to another; it may be the typical (and we say 'How true!') or the abnormal (and we say 'How strange!'); it may be the beautiful, the terrible, the awe-inspiring, the exhilarating, the pathetic, the comic, or the merely piquant. Literature gives the entrée to them all.

Those of us who have been true readers all our life seldom fully realise the enormous extension of our being which we owe to authors. We realise it best when we talk with an unliterary friend. He may be full of goodness and good sense but he inhabits a tiny world. In it, we should be suffocated. The man who is contented to be only himself, and therefore less a self, is in prison. My own eyes are not enough for me, I will see through those of others. Reality, even seen through the eyes of many, is not enough. I will see what others have invented. Even the eyes of all humanity are not enough. I regret that the brutes can-

not write books. Very gladly would I learn what face things present to a mouse or a bee; more gladly still would I perceive the olfactory world charged with all the information and emotion it carries for a dog.

Literary experience heals the wound, without undermining the privilege, of individuality. There are mass emotions which heal the wound; but they destroy the privilege. In them our separate selves are pooled and we sink back into sub-individuality. But in reading great literature I become a thousand men and yet remain myself. Like the night sky in the Greek poem, I see with a myriad eyes, but it is still I who see. Here, as in worship, in love, in moral action, and in knowing, I transcend myself; and am never more myself than when I do.

HOW TO KNOW IF YOU ARE A TRUE READER

1. *Loves to re-read books.*

The majority never read anything twice. The sure mark of an unliterary man is that he considers 'I've read it already' to be a conclusive argument against reading a work. We have all known women who remembered a novel so dimly that they had to stand for half an hour in the library skimming through it before they were certain they had once read it. But the moment they became certain, they rejected it immediately. It was for them dead, like a burnt-out

---

An Experiment in Criticism
(from Chapter 1, "A Few and the Many")

match, an old railway ticket, or yesterday's paper; they had already used it. Those who read great works, on the other hand, will read the same work ten, twenty or thirty times during the course of their life.

2. *Highly values reading as an activity (versus as a last resort).*

Secondly, the majority, though they are sometimes frequent readers, do not set much store by reading. They turn to it as a last resource. They abandon it with alacrity as soon as any alternative pastime turns up. It is kept for railway journeys, illnesses, odd moments of enforced solitude, or for the process called 'reading oneself to sleep'. They sometimes combine it with desultory conversation; often, with listening to the radio. But literary people are always looking for leisure and silence in which to read and do so with their whole attention. When they are denied

such attentive and undisturbed reading even for a few days they feel impoverished.

3. *Lists the reading of particular books as a life-changing experience.*

Thirdly, the first reading of some literary work is often, to the literary, an experience so momentous that only experiences of love, religion, or bereavement can furnish a standard of comparison. Their whole consciousness is changed. They have become what they were not before. But there is no sign of anything like this among the other sort of readers. When they have finished the story or the novel, nothing much, or nothing at all, seems to have happened to them.

4. *Continuously reflects and recalls what one has read.*

Finally, and as a natural result of their different behaviour in reading, what they have read is constantly

and prominently present to the mind of the few, but not to that of the many. The former mouth over their favourite lines and stanzas in solitude. Scenes and characters from books provide them with a sort of iconography by which they interpret or sum up their own experience. They talk to one another about books, often and at length. The latter seldom think or talk of their reading.

It is pretty clear that the majority, if they spoke without passion and were fully articulate, would not accuse us of liking the wrong books, but of making such a fuss about any books at all. We treat as a main ingredient in our well-being something which to them is marginal. Hence to say simply that they like one thing and we another is to leave out nearly the whole of the facts. If like is the correct word for what they do to books, some other word must be found for what we do. Or, conversely, if we like our kind of book we must not say that they like any book. If the few have 'good taste', then we may have to say that no such thing as 'bad taste' exists: for the inclination which the many have to their sort of reading is not the same thing and, if the word were univocally used, would not be called taste at all. . . .

Many people enjoy popular music in a way which is compatible with humming the tune, stamping in time, talking, and eating. And when the popular tune has once gone out of fashion they enjoy it no more. Those who enjoy Bach react quite differently. Some buy pictures because the walls 'look so bare without them'; and after the pictures have been in the house for a week they become practically invisible to them. But there are a few who feed on a great picture for years. As regards nature, the majority 'like a nice view as well as anyone'. They are not saying a word against it. But to make the landscapes a really important factor in, say, choosing the place for a holiday—to put them on a level with such serious considerations as a luxurious hotel, a good golf links, and a sunny climate—would seem to them affectation.

WHY CHILDREN'S STORIES ARE NOT JUST FOR CHILDREN

I AM ALMOST INCLINED TO SET IT UP AS A CANON THAT a children's story which is enjoyed only by children is a bad children's story. The good ones last. A waltz which you can like only when you are waltzing is a bad waltz.

This canon seems to me most obviously true of that particular type of children's story which is dearest to my own taste, the fantasy or fairy tale. Now the modern critical world uses 'adult' as a term of

---

Of Other Worlds
(from "On Three Ways of Writing for Children")

approval. It is hostile to what it calls 'nostalgia' and contemptuous of what it calls 'Peter Pantheism'. Hence a man who admits that dwarfs and giants and talking beasts and witches are still dear to him in his fifty-third year is now less likely to be praised for his perennial youth than scorned and pitied for arrested development. If I spend some little time defending myself against these charges, this is not so much because it matters greatly whether I am scorned and pitied as because the defence is germane to my whole view of the fairy tale and even of literature in general. My defence consists of three propositions.

(1) I reply with a *tu quoque* ["you also"]. Critics who treat *adult* as a term of approval, instead of as a merely descriptive term, cannot be adult themselves. To be concerned about being grown up, to admire the grown up because it is grown up, to blush at the suspicion of being childish; these things are the marks of childhood and adolescence. And in childhood and

adolescence they are, in moderation, healthy symptoms. Young things ought to want to grow. But to carry on into middle life or even into early manhood this concern about being adult is a mark of really arrested development. When I was ten, I read fairy tales in secret and would have been ashamed if I had been found doing so. Now that I am fifty I read them openly. When I became a man I put away childish things, including the fear of childishness and the desire to be very grown up.

(2) The modern view seems to me to involve a false conception of growth. They accuse us of arrested development because we have not lost a taste we had in childhood. But surely arrested development consists not in refusing to lose old things but in failing to add new things? I now like hock, which I am sure I should not have liked as a child. But I still like lemon-squash. I call this growth or development because I have been enriched: where I formerly

had only one pleasure, I now have two. But if I had to lose the taste for lemon-squash before I acquired the taste for hock, that would not be growth but simple change. I now enjoy Tolstoy and Jane Austen and Trollope as well as fairy tales and I call that growth: if I had had to lose the fairy tales in order to acquire the novelists, I would not say that I had grown but only that I had changed. A tree grows because it adds rings: a train doesn't grow by leaving one station behind and puffing on to the next. In reality, the case is stronger and more complicated than this. I think my growth is just as apparent when I now read the fairy tales as when I read the novelists, for I now enjoy the fairy tales better than I did in childhood: being now able to put more in, of course I get more out. But I do not here stress that point. Even if it were merely a taste for grown-up literature added to an unchanged taste for children's literature, addition would still be entitled to the name 'growth', and the process of merely dropping one parcel when you pick up another would not. It is, of course, true that the process of growing does, incidentally and unfortunately, involve some more losses. But that is not the essence of growth, certainly not what makes growth admirable or desirable. If it were, if to drop parcels and to leave stations behind were the essence and virtue of growth, why should we stop at the adult? Why should not *senile* be equally a term of approval? Why are we not to be congratulated on losing our teeth and hair? Some critics seem to confuse growth with the cost of growth and also to wish to make that cost far higher than, in nature, it need be.

(3) The whole association of fairy tale and fantasy with childhood is local and accidental. I hope everyone has read Tolkien's essay on fairy tales, which is perhaps the most important contribution to the subject that anyone has yet made. If so, you will

know already that, in most places and times, the fairy tale has not been specially made for, nor exclusively enjoyed by, children. It has gravitated to the nursery when it became unfashionable in literary circles, just as unfashionable furniture gravitated to the nursery in Victorian houses. In fact, many children do not like this kind of book, just as many children do not like horsehair sofas: and many adults do like it, just as many adults like rocking chairs. And those who do like it, whether young or old, probably like it for the same reason. And none of us can say with any certainty what that reason is. The two theories which are most often in my mind are those of Tolkien and of Jung.

According to Tolkien[1] the appeal of the fairy story lies in the fact that man most fully exercises his function as a 'subcreator'; not, as they love to say

1 J. R. R. Tolkien, 'On Fairy-Stories', *Essays Presented to Charles Williams* (1947), p. 66 ff.

now, making a 'comment upon life' but making, so far as possible, a subordinate world of his own. Since, in Tolkien's view, this is one of man's proper functions, delight naturally arises whenever it is successfully performed. For Jung, fairy tale liberates Archetypes which dwell in the collective unconscious, and when we read a good fairy tale we are obeying the old precept 'Know thyself'. I would venture to add to this my own theory, not indeed of the Kind as a whole, but of one feature in it: I mean, the presence of beings other than human which yet behave, in varying degrees, humanly: the giants and dwarfs and talking beasts. I believe these to be at least (for they may have many other sources of power and beauty) an admirable hieroglyphic which conveys psychology, types of character, more briefly than novelistic presentation and to readers whom novelistic presentation could not yet reach. Consider Mr Badger in *The Wind in the Willows*—that

When I was ten, I read fairy tales in secret and would have been ashamed if I had been found doing so. Now that I am fifty I read them openly. When I became a man I put away childish things, including the fear of childishness and the desire to be very grown up.

extraordinary amalgam of high rank, coarse manners, gruffness, shyness, and goodness. The child who has once met Mr Badger has ever afterwards, in its bones, a knowledge of humanity and of English social history which it could not get in any other way.

Of course as all children's literature is not fantastic, so all fantastic books need not be children's books. It is still possible, even in an age so ferociously anti-romantic as our own, to write fantastic stories for adults: though you will usually need to have made a name in some more fashionable kind of literature before anyone will publish them. But there may be an author who at a particular moment finds not only fantasy but fantasy-for-children the exactly right form for what he wants to say. The distinction is a fine one. His fantasies for children and his fantasies for adults will have very much more in common with one another than either has with the ordinary novel or with what is sometimes called 'the novel of child life'. Indeed the same readers will probably read both his fantastic 'juveniles' and his fantastic stories for adults. For I need not remind such an audience as this that the neat sorting-out of books into age-groups, so dear to publishers, has only a very sketchy relation with the habits of any real readers. Those of us who are blamed when old for reading childish books were blamed when children for reading books too old for us. No reader worth his salt trots along in obedience to a time-table.

LITERATURE AS TIME TRAVEL

MANY WILL THINK IT REASONABLE TO EXAMINE children in Geography or (Heaven help us!) in Divinity, yet not in English, on the ground that Geography and Divinity were never intended to entertain, whereas Literature was. The teaching of English Literature, in fact, is conceived simply as an aid to 'appreciation'. And appreciation is, to be sure, a sine qua non. To have laughed at the jokes, shuddered at the tragedy, wept at the pathos—this is as

---

Present Concerns
(from "The Death of English")

necessary as to have learned grammar. But neither grammar nor appreciation is the ultimate End.

The true aim of literary studies is to lift the student out of his provincialism by making him 'the spectator', if not of all, yet of much, 'time and existence'. The student, or even the schoolboy, who has been brought by good (and therefore mutually disagreeing) teachers to meet the past where alone the past still lives, is taken out of the narrowness of his own age and class into a more public world. He is learning the true *Phaenomenologie des Geistes*; discovering what varieties there are in Man.

'History' alone will not do, for it studies the past mainly in secondary authorities. It is possible to 'do History' for years without knowing at the end what it felt like to be an Anglo Saxon eorl, a cavalier, an eighteenth-century country gentleman. The gold behind the paper currency is to be found, almost exclusively, in literature. In it lies deliverance from the tyranny of generalisations and catchwords. Its students know (for example) what diverse realities—Launcelot, Baron Bradwardine, Mulvaney—hide behind the word militarism.

The true aim of literary studies is to lift the student out of his provincialism by making him 'the spectator', if not of all, yet of much, 'time and existence'.

If I regard the English Faculties at our Universities as the chief guardians (under modern conditions) of the Humanities, I may doubtless be misled by partiality for studies to which I owe so much; yet in a way I am well placed for judging. I have been pupil and teacher alike in Literae Humaniores, pupil and teacher alike in English; in the History School (I confess) teacher only. If anyone said that English was now the most liberal—and liberating—discipline of the three, I should not find it easy to contradict him.

WHY FAIRY TALES ARE OFTEN LESS DECEPTIVE THAN 'REALISTIC' STORIES

About once every hundred years some wiseacre gets up and tries to banish the fairy tale. Perhaps I had better say a few words in its defence, as reading for children.

It is accused of giving children a false impression of the world they live in. But I think no literature that children could read gives them less of a false impression. I think what profess to be realistic stories

---

Of Other Worlds
(from "On Three Ways of Writing for Children")

for children are far more likely to deceive them. I never expected the real world to be like the fairy tales. I think that I did expect school to be like the school stories. The fantasies did not deceive me: the school stories did. All stories in which children have adventures and successes which are possible, in the sense that they do not break the laws of nature, but almost infinitely improbable, are in more danger than the fairy tales of raising false expectations.

Almost the same answer serves for the popular charge of escapism, though here the question is not so simple. Do fairy tales teach children to retreat into a world of wish-fulfilment—'fantasy' in the technical psychological sense of the word—instead of facing the problems of the real world? Now it is here that the problem becomes subtle. Let us again lay the fairy tale side by side with the school story or any other story which is labelled a 'Boy's Book' or a 'Girl's Book', as distinct from a 'Children's Book'.

There is no doubt that both arouse, and imaginatively satisfy, wishes. We long to go through the looking glass, to reach fairy land. We also long to be the immensely popular and successful schoolboy or schoolgirl, or the lucky boy or girl who discovers the spy's plot or rides the horse that none of the cowboys can manage. But the two longings are very different. The second, especially when directed on something so close as school life, is ravenous and deadly serious. Its fulfilment on the level of imagination is in very truth compensatory: we run to it from the disappointments and humiliations of the real world: it sends us back to the real world undivinely discontented. For it is all flattery to the ego. The pleasure consists in picturing oneself the object of admiration. The other longing, that for fairy land, is very different. In a sense a child does not long for fairy land as a boy longs to be the hero of the first eleven [the top class team of a school in either football/soccer or in cricket]. Does anyone suppose that he really and prosaically longs for all the dangers and discomforts of a fairy tale?—really wants dragons in contemporary England? It is not so. It would be much truer to

[A child] does not despise real woods because he has read of enchanted woods: the reading makes all real woods a little enchanted.

say that fairy land arouses a longing for he knows not what. It stirs and troubles him (to his life-long enrichment) with the dim sense of something beyond his reach and, far from dulling or emptying the

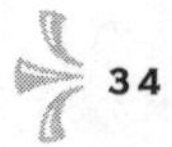

actual world, gives it a new dimension of depth. He does not despise real woods because he has read of enchanted woods: the reading makes all real woods a little enchanted. This is a special kind of longing. The boy reading the school story of the type I have in mind desires success and is unhappy (once the book is over) because he can't get it: the boy reading the fairy tale desires and is happy in the very fact of desiring. For his mind has not been concentrated on himself, as it often is in the more realistic story.

I do not mean that school stories for boys and girls ought not to be written. I am only saying that they are far more liable to become 'fantasies' in the clinical sense than fantastic stories are. And this distinction holds for adult reading too. The dangerous fantasy is always superficially realistic. The real victim of wishful reverie does not batten on the *Odyssey*, *The Tempest*, or *The Worm Ouroboros*: he (or she) prefers stories about millionaires, irresistible beauties, posh hotels, palm beaches and bedroom scenes—things that really might happen, that ought to happen, that would have happened if the reader had had a fair chance. For, as I say, there are two kinds of longing. The one is an *askesis*, a spiritual exercise, and the other is a disease.

A far more serious attack on the fairy tale as children's literature comes from those who do not wish children to be frightened. I suffered too much from night-fears myself in childhood to undervalue this objection. I would not wish to heat the fires of that private hell for any child. On the other hand, none of my fears came from fairy tales. Giant insects were my specialty, with ghosts a bad second. I suppose the ghosts came directly or indirectly from stories, though certainly not from fairy stories, but I don't think the insects did. I don't know anything my parents could have done or left undone which would have saved me from the pincers, mandibles, and eyes of those many-legged abominations. And that, as so many people have pointed out, is the difficulty. We do not know what will or will not frighten a child in this particular way. I say 'in this particular way' for we must here make a distinction. Those who say that children must not be frightened may mean two things. They may mean (1) that we must not do anything likely to give the child those haunting, disabling, pathological fears against which ordinary courage is helpless: in fact, *phobias*. His mind must, if possible, be kept clear of things he can't bear to think of. Or they may mean (2) that we must try to keep out of his mind the knowledge that he is born into a world of death, violence, wounds, adventure, heroism and cowardice, good and evil. If they mean the first I agree with them: but not if they mean the second. The second would indeed be to give children a false impression and feed them on escapism in the bad sense.

There is something ludicrous in the idea of so educating a generation which is born to the Ogpu and the atomic bomb. Since it is so likely that they will meet cruel enemies, let them at least have heard of brave knights and heroic courage. Otherwise you are making their destiny not brighter but darker. Nor do most of us find that violence and bloodshed, in a story, produce any haunting dread in the minds of children. As far as that goes, I side impenitently with the human race against the modern reformer. Let there be wicked kings and beheadings, battles and dungeons, giants and dragons, and let villains be soundly killed at the end of the book. Nothing will persuade me that this causes an ordinary child any kind or degree of fear beyond what it wants, and needs, to feel. For, of course, it wants to be a little frightened.

The other fears—the phobias—are a different matter. I do not believe one can control them by literary means. We seem to bring them into the world with us ready made. No doubt the particular image on which the child's terror is fixed can sometimes be traced to a book. But is that the source, or only the occasion, of the fear? If he had been spared that image, would not some other, quite unpredictable by you, have had the same effect? Chesterton has told us of a boy who was more afraid of the Albert Memorial than anything else in the world. I know a man whose great childhood terror was the India paper edition of the *Encyclopaedia Britannica*—for a reason I defy you to guess. And I think it possible that by confining your child to blameless stories of child life in which nothing at all alarming ever happens, you would fail to banish the terrors, and would succeed in banishing all that can ennoble them or make them endurable. For in the fairy tales, side by side with the terrible figures, we find the immemorial comforters and protectors, the radiant ones; and the terrible figures are not merely terrible, but sublime. It would be nice if no little boy in bed, hearing, or thinking he hears, a sound, were ever at all frightened. But if he is going to be frightened, I think it better that he should think of giants and dragons than merely of burglars. And I think St George, or any bright champion in armour, is a better comfort than the idea of the police.

I will even go further. If I could have escaped all my own night-fears at the price of never having known 'faerie', would I now be the gainer by that bargain? I am not speaking carelessly. The fears were very bad. But I think the price would have been too high.

THE CASE FOR READING OLD BOOKS

THERE IS A STRANGE IDEA ABROAD THAT IN EVERY subject the ancient books should be read only by the professionals, and that the amateur should content himself with the modern books. Thus I have found as a tutor in English Literature that if the average student wants to find out something about Platonism, the very last thing he thinks of doing is to take a translation of Plato off the library shelf and read the Symposium. He would rather read some dreary modern book ten times as long, all about

---

God in the Dock
(from "On the Reading of Old Books")

'isms' and influences and only once in twelve pages telling him what Plato actually said. The error is rather an amiable one, for it springs from humility. The student is half afraid to meet one of the great philosophers face to face. He feels himself inadequate and thinks he will not understand him. But if he only knew, the great man, just because of his greatness, is much more intelligible than his modern commentator. The simplest student will be able to understand, if not all, yet a very great deal of what Plato said; but hardly anyone can understand some modern books on Platonism. It has always therefore been one of my main endeavours as a teacher to persuade the young that firsthand knowledge is not only more worth acquiring than secondhand knowledge, but is usually much easier and more delightful to acquire.

This mistaken preference for the modern books and this shyness of the old ones is nowhere more rampant than in theology. Wherever you find a little study circle of Christian laity you can be almost certain that they are studying not St Luke or St Paul or St Augustine or Thomas Aquinas or Hooker or Butler, but M. Berdyaev or M. Maritain or M. Niebuhr or Miss Sayers or even myself.

Now this seems to me topsy-turvy. Naturally, since I myself am a writer, I do not wish the ordinary reader to read no modern books. But if he must read only the new or only the old, I would advise him to read the old. And I would give him this advice precisely because he is an amateur and therefore much less protected than the expert against the dangers of an exclusive contemporary diet. A new book is still on its trial and the amateur is not in a position to judge it. It has to be tested against the great body of Christian thought down the ages, and all its hidden implications (often unsuspected by the author himself) have to be brought to light. Often it cannot be fully understood without the knowledge of a

Every age has its own outlook. It is specially good at seeing certain truths and specially liable to make certain mistakes. We all, therefore, need the books that will correct the characteristic mistakes of our own period. And that means the old books.

good many other modern books. If you join at eleven o'clock a conversation which began at eight you will often not see the real bearing of what is said. Remarks which seem to you very ordinary will produce laughter or irritation and you will not see why—the reason, of course, being that the earlier stages of the conversation have given them a special point. In the same way sentences in a modern book which look quite ordinary may be directed at some other book; in this way you may be led to accept what you would have indignantly rejected if you knew its real significance. The only safety is to have a standard of plain, central Christianity ('mere Christianity' as Baxter called it) which puts the controversies of the moment in their proper perspective. Such a standard can be acquired only from the old books. It is a good rule, after reading a new book, never to allow yourself another new one till you have read an old one in between. If that is too much for you, you should at least read one old one to every three new ones.

Every age has its own outlook. It is specially good at seeing certain truths and specially liable to make certain mistakes. We all, therefore, need the books that will correct the characteristic mistakes of our own period. And that means the old books. All contemporary writers share to some extent the contemporary outlook—even those, like myself, who seem most opposed to it. Nothing strikes me more when I read the controversies of past ages than the fact that both sides were usually assuming without question a good deal which we should now absolutely deny. They thought that they were as completely opposed as two sides could be, but in fact they were all the time secretly united—united with each other and against earlier and later ages—by a great mass of common assumptions.

We may be sure that the characteristic blindness of the twentieth century—the blindness about which posterity will ask, 'But how could they have thought that?'—lies where we have never suspected it, and concerns something about which there is untroubled agreement between Hitler and President Roosevelt or between Mr. H. G. Wells and Karl Barth. None of us can fully escape this blindness, but we shall certainly increase it, and weaken our guard against it, if we read only modern books. Where they are true they will give us truths which we half knew already. Where they are false they will aggravate the error with which we are already dangerously ill. The only palliative is to keep the clean sea breeze of the centuries blowing through our minds, and this can be done only by reading old books. Not, of course, that there is any magic about the past. People were no cleverer then than they are now; they made as many mistakes as we. But not the same mistakes. They will not flatter us in the errors we are already committing; and their own errors, being now open and palpable, will not endanger us. Two heads are better than one, not because either is infallible, but because they are unlikely to go wrong in the same direction. To be sure, the books of the future would be just as good a corrective as the books of the past, but unfortunately we cannot get at them.

ON THE ROLE OF THE MARVELLOUS

GOOD STORIES OFTEN INTRODUCE THE MARVELLOUS or supernatural, and nothing about Story has been so often misunderstood as this. Thus, for example, Dr Johnson, if I remember rightly, thought that children liked stories of the marvellous because they were too ignorant to know that they were impossible. But children do not always like them, nor are those who like them always children; and to enjoy reading about fairies—much more about giants and dragons—it is not necessary to believe in

---

Of Other Worlds
(from "On Stories")

them. Belief is at best irrelevant; it may be a positive disadvantage. Nor are the marvels in good Story ever mere arbitrary fictions stuck on to make the narrative more sensational. I happened to remark to a man who was sitting beside me at dinner the other night that I was reading Grimm in German of an evening but never bothered to look up a word I didn't know, 'so that it is often great fun' (I added) 'guessing what it was that the old woman gave to the prince which he afterwards lost in the wood'. 'And specially difficult in a fairy-tale,' said he, 'where everything is arbitrary and therefore the object might be anything at all.' His error was profound. The logic of a fairy tale is as strict as that of a realistic novel, though different.

Does anyone believe that Kenneth Grahame made an arbitrary choice when he gave his principal character the form of a toad, or that a stag, a pigeon, a lion, would have done as well? The choice is based on the fact that the real toad's face has a grotesque resemblance to a certain kind of human face—a rather apoplectic face with a fatuous grin on it. This is, no doubt, an accident in the sense that all the lines which suggest the resemblance are really there for quite different biological reasons. The ludicrous quasi-human expression is therefore changeless: the toad cannot stop grinning because its 'grin' is not really a grin at all. Looking at the creature we thus see, isolated and fixed, an aspect of human vanity in its funniest and most pardonable form; following that hint Grahame creates Mr Toad—an ultra-Jonsonian 'humour'. And we bring back the wealth of the Indies; we have henceforward more amusement in, and kindness towards, a certain kind of vanity in real life.

But why should the characters be disguised as animals at all? The disguise is very thin, so thin that Grahame makes Mr Toad on one occasion 'comb the dry leaves out of his *hair*'. Yet it is quite indispensable. If you try to rewrite the book with all the characters humanized you are faced at the outset with a dilemma. Are they to be adults or children? You will find that they can be neither. They are like children in so far as they have no responsibilities, no struggle for existence, no domestic cares. Meals turn up; one does not even ask who cooked them. In Mr Badger's kitchen 'plates on the dresser grinned at pots on the shelf'. Who kept them clean? Where were they bought? How were they delivered in the Wild Wood? Mole is very snug in his subterranean home, but what was he living *on*? If he is a *rentier* where is the bank, what are his investments? The tables in his forecourt were 'marked with rings that hinted at beer mugs'. But where did he get the beer? In that way the life of all the characters is that of children for whom everything is provided and who take everything for granted. But in other ways it is

No book is really worth reading at the age of ten which is not equally (and often far more) worth reading at the age of fifty. . . . The only imaginative works we ought to grow out of are those which it would have been better not to have read at all.

the life of adults. They go where they like and do what they please, they arrange their own lives.

To that extent the book is a specimen of the most scandalous escapism: it paints a happiness under incompatible conditions—the sort of freedom we can have only in childhood and the sort we can have only in maturity—and conceals the contradiction by the further pretence that the characters are not human beings at all. The one absurdity helps to hide the other. It might be expected that such a book would unfit us for the harshness of reality and send us back to our daily lives unsettled and discontented. I do not find that it does so. The happiness which it presents to us is in fact full of the simplest and most attainable things—food, sleep, exercise, friendship, the face of nature, even (in a sense) religion. That 'simple but sustaining meal' of 'bacon and broad beans and a macaroni pudding' which Rat gave to his friends has, I doubt not, helped down many a real nursery dinner. And in the same way the whole story, paradoxically enough, strengthens our relish for real life. This excursion into the preposterous sends us back with renewed pleasure to the actual.

It is usual to speak in a playfully apologetic tone about one's adult enjoyment of what are called 'children's books'. I think the convention a silly one. No book is really worth reading at the age of ten which is not equally (and often far more) worth reading at the age of fifty—except, of course, books of information. The only imaginative works we ought to grow out of are those which it would have been better not to have read at all. A mature palate will probably not much care for *crème de menthe*: but it ought still to enjoy bread and butter and honey.

GROWING UP AMIDST A SEA OF BOOKS

1. *A Product of Endless Books*

The New House [Little Lea, Lewis's childhood home] is almost a major character in my story. I am a product of long corridors, empty sunlit rooms, upstairs indoor silences, attics explored in solitude, distant noises of gurgling cisterns and pipes, and the noise of wind under the tiles. Also, of endless books. My father bought all the books he read and never got rid of any of them. There were books in the study,

---

Surprised by Joy
(from "The First Years")

books in the drawing room, books in the cloakroom, books (two deep) in the great bookcase on the landing, books in a bedroom, books piled as high as my shoulder in the cistern attic, books of all kinds reflecting every transient stage of my parents' interest, books readable and unreadable, books suitable for a child and books most emphatically not. Nothing was forbidden me. In the seemingly endless rainy afternoons I took volume after volume from the shelves. I had always the same certainty of finding a book that was new to me as a man who walks into a field has of finding a new blade of grass.

II. *The Ideal Day for a Young Scholar*

[At Great Bookham] We now settled into a routine which has ever since served in my mind as an archetype, so that what I still mean when I speak of a 'normal' day (and lament that normal days are so rare) is a day of the Bookham pattern. For if I could please myself I would always live as I lived there. I would choose always to breakfast at exactly eight and to be at my desk by nine, there to read or write till one. If a cup of good tea or coffee could be brought me about eleven, so much the better. A step or so out of doors for a pint of beer would not do

---

Surprised by Joy
(from "The Great Knock")

quite so well; for a man does not want to drink alone and if you meet a friend in the taproom the break is likely to be extended beyond its ten minutes. At one precisely lunch should be on the table; and by two at the latest I would be on the road. Not, except at rare intervals, with a friend. Walking and talking are two very great pleasures, but it is a mistake to combine them. Our own noise blots out the sounds and silences of the outdoor world; and talking leads almost inevitably to smoking, and then farewell to nature as far as one of our senses is concerned. The only friend to walk with is one (such as I found, during the holidays, in Arthur) who so exactly shares your taste for each mood of the countryside that a glance, a halt, or at most a nudge, is enough to assure us that the pleasure is shared. The return from the walk, and the arrival of tea, should be exactly coincident, and not later than a quarter past four. Tea should be taken in solitude, as I took it at Bookham on those (happily numerous) occasions when Mrs Kirkpatrick was out; the Knock himself disdained this meal. For eating and reading are two pleasures that combine admirably. Of course not all books are suitable for mealtime reading. It would be a kind of blasphemy to read poetry at table. What one wants is a gossipy, formless book which can be opened anywhere. The ones I learned so to use at Bookham were Boswell, and a translation of Herodotus, and Lang's *History of English Literature*. *Tristram Shandy*, *Elia* and the *Anatomy of Melancholy* are all good for the same purpose. At five a man should be at work again, and at it till seven. Then, at the evening meal and after, comes the time for talk, or, failing that, for lighter reading; and unless you are making a night of it with your cronies (and at Bookham I had none) there is no reason why you should ever be in bed later than eleven. But when is a man to write his letters? You forget that I am describing the happy life I led with Kirk or the ideal life I would live now if I could. And it is an essential of the happy life that a man would have almost no mail and never dread the postman's knock.

III. *Learning to Love the Bodies of Books*

One other thing that Arthur [Greeves] taught me was to love the bodies of books. I had always respected them. My brother and I might cut up stepladders without scruple; to have thumb-marked or dog's-eared a book would have filled us with shame. But Arthur did not merely respect, he was enamored; and soon, I too. The setup of the page, the feel and smell of the paper, the differing sounds that different papers make as you turn the leaves, became sensuous delights. This revealed to me a flaw in

---

Surprised by Joy
(from "Fortune's Smile")

Kirk. How often have I shuddered when he took a new classical text of mine in his gardener's hands, bent back the boards till they creaked, and left his sign on every page.

'Yes I remember,' said my father. 'That was old Knock's one fault.'

'A bad one,' said I.

'An all but unforgivable one,' said my father.

ON ENCOUNTERING A FAVORITE AUTHOR FOR THE FIRST TIME

I. *Siren Songs and the Wind of Joy*

The glorious week-end of reading was before me. Turning to the bookstall, I picked out an Everyman in a dirty jacket, *Phantastes, a faerie Romance*, George MacDonald. . . . That evening I began to read my new book.

The woodland journeyings in that story, the ghostly enemies, the ladies both good and evil, were

---

Surprised by Joy
(from "Check")

close enough to my habitual imagery to lure me on without the perception of a change. It is as if I were carried sleeping across the frontier, or as if I had died in the old country and could never remember how I came alive in the new. For in one sense the new country was exactly like the old. I met there all that had already charmed me in Malory, Spenser, Morris, and Yeats. But in another sense all was changed. I did not yet know (and I was long in learning) the name of the new quality, the bright shadow, that rested on the travels of Anodos [the protagonist in *Phantastes*]. I do now. It was Holiness. For the first time the song of the sirens sounded like the voice of my mother or my nurse. Here were old wives' tales; there was nothing to be proud of in enjoying them. It was as though the voice which had called to me from the world's end were now speaking at my side. It was with me in the room, or in my own body, or behind me. If it had once eluded me by its distance, it now eluded me by proximity—

something too near to see, too plain to be understood, on this side of knowledge. It seemed to have been always with me; if I could ever have turned my head quick enough I should have seized it. Now for the first time I felt that it was out of reach not because of something I could not do but because of something I could not stop doing. If I could only leave off, let go, unmake myself, it would be there. Meanwhile, in this new region all the confusions that had hitherto perplexed my search for Joy were disarmed. There was no temptation to confuse the scenes of the tale with the light that rested upon them, or to suppose that they were put forward as realities, or even to dream that if they had been realities and I could reach the woods where Anodos journeyed I should thereby come a step nearer to my desire. Yet, at the same time, never had the wind of Joy blowing through any story been less separable from the story itself. Where the god and the *idolon* were most nearly one there was least danger of confounding them. Thus, when the great moments came I did not break away from the woods and cottages that I read of to seek some bodiless light shining beyond them, but gradually, with a swelling continuity (like the sun at

> ***It was as though the voice which had called to me from the world's end were now speaking at my side.***

mid-morning burning through a fog) I found the light shining on those woods and cottages, and then on my own past life, and on the quiet room where I sat and on my old teacher where he nodded above his little *Tacitus*. For I now perceived that while the air of the new region made all my erotic and magical perversions of Joy look like sordid trumpery, it had no such disen-

chanting power over the bread upon the table or the coals in the grate. That was the marvel. Up till now each visitation of Joy had left the common world momentarily a desert—'The first touch of the earth went nigh to kill'. Even when real clouds or trees had been the material of the vision, they had been so only by reminding me of another world; and I did not like the return to ours. But now I saw the bright shadow coming out of the book into the real world and resting there, transforming all common things and yet itself unchanged. Or, more accurately, I saw the common things drawn into the bright shadow. *Unde hoc mihi?* [Luke 1:43, Vulgate Latin]. In the depth of my disgraces, in the then invincible ignorance of my intellect, all this was given me without asking, even without consent. That night my imagination was, in a certain sense, baptized; the rest of me, not unnaturally, took longer. I had not the faintest notion what I had let myself in for by buying *Phantastes*.

II. *Crossing a Great Frontier*

I have never concealed the fact that I regarded [George MacDonald] as my master; indeed I fancy I have never written a book in which I did not quote from him. But it has not seemed to me that those who have received my books kindly take even now sufficient notice of the affiliation. Honesty drives me to emphasize it. And even if honesty did not—well, I am a don, and 'source-hunting' (*Quellenforschung*) is perhaps in my marrow. It must be more than thirty years ago that I bought—almost unwillingly, for I had looked at the volume on that bookstall and re-

---

George MacDonald: An Anthology
(from the Preface)

jected it on a dozen previous occasions—the Everyman edition of *Phantastes*. A few hours later I knew that I had crossed a great frontier.

I had already been waist-deep in romanticism; and likely enough, at any moment, to founder into its darker and more evil forms, slithering down the steep descent that leads from the love of strangeness to that of eccentricity and thence to that of perversity. Now *Phantastes* was romantic enough in all conscience; but there was a difference. Nothing was at that time further from my thoughts than Christianity and I therefore had no notion what this difference really was. I was only aware that if this new world was strange, it was also homely and humble; that if this was a dream, it was a dream in which one at least felt strangely vigilant; that the whole book had about it a sort of cool, morning innocence, and also, quite unmistakably, a certain quality of Death, *good* Death.

What it actually did to me was to convert, even to baptize (that was where the Death came in) my imagination. It did nothing to my intellect nor (at that time) to my conscience. Their turn came far later and with the help of many other books and men. But when the process was complete—by which, of course, I mean 'when it had *really* begun'—I found that I was still with MacDonald and that he had accompanied me all the way and that I was now at last ready to hear from him much that he could not have told me at that first meeting. But in a sense, what he was now telling me was the very same that he had told me from the beginning. There was no question of getting through to the kernel and throwing away the shell: no question of a gilded pill. The pill was gold all through. The quality which had enchanted me in his imaginative works turned out to be the quality of the real universe, the divine, magical, terrifying, and ecstatic reality in which we all live.

WHY MOVIES SOMETIMES RUIN BOOKS

I WAS ONCE TAKEN TO SEE A FILM VERSION OF *KING SOLOMON'S MINES*. Of its many sins—not least the introduction of a totally irrelevant young woman in shorts who accompanied the three adventurers wherever they went—only one here concerns us. At the end of Haggard's book, as everyone remembers, the heroes are awaiting death entombed in a rock chamber and surrounded by the mummified kings of that land. The maker of the film version, however, apparently thought this tame. He substituted a

Of Other Worlds
(from "On Stories")

subterranean volcanic eruption, and then went one better by adding an earthquake. Perhaps we should not blame him. Perhaps the scene in the original was not 'cinematic' and the man was right, by the canons of his own art, in altering it. But it would have been better not to have chosen in the first place a story which could be adapted to the screen only by being ruined. Ruined, at least, for me.

No doubt if sheer excitement is all you want from a story, and if increase of dangers increases excitement, then a rapidly changing series of two risks (that of being burned alive and that of being crushed to bits) would be better than the single prolonged danger of starving to death in a cave. But that is just the point.

There must be a pleasure in such stories distinct from mere excitement or I should not feel that I had been cheated in being given the earthquake instead of Haggard's actual scene. What I lose is the whole sense of the deathly (quite a different thing from simple danger of death)—the cold, the silence, and the surrounding faces of the ancient, the crowned and sceptred, dead. You may, if you please, say that Rider Haggard's effect is quite as 'crude' or 'vulgar' or 'sensational' as that which the film substituted for it. I am not at present discussing that. The point is that it is extremely different. The one lays a hushing spell on the imagination; the other excites a rapid flutter of the nerves. In reading that chapter of the book curiosity or suspense about the escape of the heroes from their death-trap makes a very minor part of one's experience. The trap I remember for ever: how they got out I have long since forgotten.

It seems to me that in talking of books which are 'mere stories'—books, that is, which concern themselves principally with the imagined event and not with character or society—nearly everyone makes the assumption that 'excitement' is the only pleasure they ever give or are intended to give. *Excitement*, in this sense, may be defined as the alternate tension and appeasement of imagined anxiety. This is what I think untrue. In some such books, and for some readers, another factor comes in. . . .

If I am alone in this experience then, to be sure, the present essay is of merely autobiographical interest. But I am pretty sure that I am not absolutely alone. I write on the chance that some others may feel the same and in the hope that I may help them to clarify their own sensations.

In the example of *King Solomon's Mines* the producer of the film substituted at the climax one kind of danger for another and thereby, for me, ruined the story. But where excitement is the only thing that matters, kinds of danger must be irrelevant. Only degrees of danger will matter. The greater the danger and the narrower the hero's escape from it, the more exciting the story will be. But when we are con-

In talking of books which are 'mere stories—books, that is, which concern themselves principally with the imagined event and not with character or society—nearly everyone makes the assumption that excitement' is the only pleasure they ever give or are intended to give.

cerned with the 'something else' this is not so. Different kinds of danger strike different chords from the imagination. Even in real life different kinds of danger produce different kinds of fear. There may come a point at which fear is so great that such distinctions vanish, but that is another matter. There is a fear which is twin sister to awe, such as a man in war-time feels when he first comes within sound of the guns; there is a fear which is twin sister to disgust, such as a man feels on finding a snake or scorpion in his bedroom. There are taut, quivering fears (for one split second hardly distinguishable from a kind of pleasurable thrill) that a man may feel on a dangerous horse or a dangerous sea; and again, dead, squashed, flattened, numbing fears, as when we think we have cancer or cholera. There are also fears which are not of *danger* at all: like the fear of some large and hideous, though innocuous, insect or the fear of a ghost. All this, even in real life. But in imagination, where the fear does not rise to abject terror and is not discharged in action, the qualitative difference is much stronger.

I can never remember a time when it was not, however vaguely, present to my consciousness. *Jack the Giant-Killer* is not, in essence, simply the story of a clever hero surmounting danger. It is in essence the story of such a hero surmounting *danger from giants*. It is quite easy to contrive a story in which, though the enemies are of normal size, the odds against Jack are equally great. But it will be quite a different story.

HOW TO MURDER WORDS

VERBICIDE, THE MURDER OF A WORD, HAPPENS IN many ways. Inflation is one of the commonest; those who taught us to say *awfully* for 'very', *tremendous* for 'great', *sadism* for 'cruelty', and *unthinkable* for 'undesirable' were verbicides. Another way is verbiage, by which I here mean the use of a word as a promise to pay which is never going to be kept. The use of *significant* as if it were an absolute, and with no intention of ever telling us what the thing is significant of, is an example. So is *diametrically* when

---

Studies in Worlds
(from the Introduction)

The greatest cause of verbicide is the fact that most people are obviously far more anxious to express their approval and disapproval of things than to describe them.

it is used merely to put *opposite* into the superlative. Men often commit verbicide because they want to snatch a word as a party banner, to appropriate its 'selling quality'. Verbicide was committed when we exchanged *Whig* and *Tory* for *Liberal* and *Conservative*. But the greatest cause of verbicide is the fact that most people are obviously far more anxious to express their approval and disapproval of things than to describe them. Hence the tendency of words to become less descriptive and more evaluative; then to become evaluative, while still retaining some hint of the sort of goodness or badness implied; and to end up by being purely evaluative—useless synonyms for *good* or for *bad*. . . .

I am not suggesting that we can by an archaizing purism repair any of the losses that have already occurred. It may not, however, be entirely useless to resolve that we ourselves will never commit verbicide. If modern critical usage seems to be initiating a process which might finally make *adolescent* and *contemporary* mere synonyms for *bad* and *good*—and stranger things have happened—we should banish them from our vocabulary. I am tempted to adapt the couplet we see in some parks—

> Let no one say, and say it to your shame,
> That there was meaning here before you came.

SAVING WORDS FROM THE EULOGISTIC ABYSS

I THINK IT WAS MISS [ROSE] MACAULAY WHO COMplained in one of her delightful articles (strong and light as steel wire) that the dictionaries are always telling us of words 'now used only in a bad sense'; seldom or never of words 'now used only in a good sense'. It is certainly true that nearly all our terms of abuse were originally terms of description; to call a man *villain* defined his legal status long before it came to denounce his morality. The human race does not seem contented with the plain dyslo-

---

Of Other Worlds
(from "On Stories")

gistic words. Rather than say that a man is dishonest or cruel or unreliable, they insinuate that he is illegitimate, or young, or low in the social scale, or some kind of animal; that he is a 'peasant slave', a *bastard*, a *cad*, a *knave*, a *dog*, a *swine*, or (more recently) an *adolescent*.

But I doubt if that is the whole story. There are, indeed, few words which were once insulting and are now complimentary—*democrat* is the only one that comes readily to mind. But surely there are words that have become *merely* complimentary—words which once had a definable sense and which are now nothing more than noises of vague approval? The clearest example is the word *gentleman*. This was once (like *villain*) a term which defined a social and heraldic fact. The question [of] whether Snooks was a gentleman was almost as soluble as the question [of] whether he was a barrister or a Master of Arts. The same question, asked forty years ago (when it was asked very often), admitted of no solution. The word had become merely eulogistic, and the qualities on which the eulogy was based varied from moment to moment even in the mind of the same speaker. This is one of the ways in which words die. A skilful doctor of words will pronounce the disease to be mortal at that moment when the word in question begins to harbour the adjectival parasites *real* or *true*. As long as *gentleman* has a clear meaning, it is enough to say that So-and-so is a gentleman. When we begin saying that he is a 'real gentleman' or 'a true gentleman' or 'a gentleman in the truest sense' we may be sure that the word has not long to live.

I would venture, then, to enlarge Miss Macaulay's observation. The truth is not simply that words originally innocent tend to acquire a bad sense. The vocabulary of flattery and insult is continually enlarged at the expense of the vocabulary of definition.

As old horses go to the knacker's yard, or old ships to the breakers, so words in their last decay go to swell the enormous list of synonyms for *good* and *bad*. And as long as most people are more anxious to express their likes and dislikes than to describe facts, this must remain a universal truth about language.

This process is going on very rapidly at the moment. The words *abstract* and *concrete* were first coined to express a distinction which is really necessary to thought; but it is only for the very highly educated that they still do so. In popular language *concrete* now means something like 'clearly defined and practicable'; it has become a term of praise. *Abstract* (partly under the phonetic infection of *abstruse*) means 'vague, shadowy, unsubstantial'; it has become a term of reproach. *Modern*, in the mouths of many speakers, has ceased to be a chronological term; it has 'sunk into a good sense' and often means little more than 'efficient' or (in some contexts) 'kind'; '*medieval barbari-*

ties', in the mouths of the same speakers, has no reference either to the Middle Ages or to those cultures classified as barbarian. It means simply 'great or wicked cruelties'. *Conventional* can no longer be used in its proper sense without explanation. *Practical* is a mere term of approval; *contemporary*, in certain schools of literary criticism, is little better.

Men do not long continue to think what they have forgotten how to say.

To save any word from the eulogistic and dyslogistic abyss is a task worth the efforts of all who love the English language. And I can think of one word—the word *Christian*—which is at this moment on the brink. When politicians talk of '*Christian* moral standards' they are not always thinking of anything which distinguishes Christian morality from Confucian or Stoic or Benthamite morality. One often feels that it is merely one literary variant among the 'adorning epithets' which, in our political style, the expression 'moral standards' is felt to require; *civilised* (another ruined word) or *modern* or *democratic* or *enlightened* would have done just as well. But it will really be a great nuisance if the word *Christian* becomes simply a synonym for *good*. For historians, if no one else, will still sometimes need the word in its proper sense, and what will they do? That is always the trouble about allowing words to slip into the abyss. Once turn *swine* into a mere insult, and you need a new word (*pig*) when you want to talk about the animal. Once let *sadism* dwindle into a useless synonym for *cruelty*, and what do you do when you have to refer to the highly special perversion which actually afflicted M. de Sade?

It is important to notice that the danger to the word *Christian* comes not from its open enemies, but from its friends. It was not egalitarians, it was officious admirers of gentility, who killed the word *gentleman*. The other day I had occasion to say that certain people were not Christians; a critic asked how I dared say so, being unable (as of course I am) to read their hearts. I had used the word to mean 'persons who profess belief in the specific doctrines of Christianity'; my critic wanted me to use it in what he would (rightly) call 'a far deeper sense'—a sense so deep that no human observer can tell to whom it applies.

And is that deeper sense not more important? It is indeed; just as it was more important to be a 'real' gentleman than to have coat-armour. But the most important sense of a word is not always the most useful. What is the good of deepening a word's connotation if you deprive the word of all practicable denotation? Words, as well as women, can be 'killed with kindness'. And when, however reverently, you have killed a word you have also, as far as in you lay, blotted from the human mind the thing that word originally stood for. Men do not long continue to think what they have forgotten how to say.

THE ACHIEVEMENTS OF J. R. R. TOLKIEN

I. *A Review of* The Hobbit

The publishers claim that *The Hobbit*, though very unlike *Alice*, resembles it in being the work of a professor at play. A more important truth is that both belong to a very small class of books which have nothing in common save that each admits us to a world of its own—a world that seems to have been going on before we stumbled into it but which, once found by the right reader, becomes indispensable to

---

On Stories and Other Essays on Literature
(from "The Hobbit")

him. Its place is with *Alice*, *Flatland*, *Phantastes*, *The Wind in the Willows*.

To define the world of *The Hobbit* is, of course, impossible, because it is new. You cannot anticipate it before you go there, as you cannot forget it once you have gone. . . .

Prediction is dangerous: but* The Hobbit *may well prove a classic.

You must read for yourself to find out how inevitable the change is and how it keeps pace with the hero's journey. Though all is marvellous, nothing is arbitrary: all the inhabitants of Wilderland seem to have the same unquestionable right to their exis-

tence as those of our own world, though the fortunate child who meets them will have no notion—and his unlearned elders not much more—of the deep sources in our blood and tradition from which they spring.

For it must be understood that this is a children's book only in the sense that the first of many readings can be undertaken in the nursery. *Alice* is read gravely by children and with laughter by grown-ups; *The Hobbit,* on the other hand, will be funniest to its youngest readers, and only years later, at a tenth or a twentieth reading, will they begin to realise what deft scholarship and profound reflection have gone to make everything in it so ripe, so friendly, and in its own way so true. Prediction is dangerous: but *The Hobbit* may well prove a classic.

II. *A Review of* The Lord of the Rings

This book[1] is like lightning from a clear sky; as sharply different, as unpredictable in our age as *Songs of Innocence* were in theirs. To say that in it heroic romance, gorgeous, eloquent, and unashamed, has suddenly returned at a period almost pathological in its anti-romanticism is inadequate. To us, who live in that odd period, the return—and the sheer relief of it—is doubtless the important thing. But in the history of Romance itself—a history which stretches

---

On Stories and Other Essays on Literature
(from "Tolkien's *The Lord of the Rings*")

---

1 *The Fellowship of the Ring* (1954), the first volume of the trilogy *The Lord of the Rings*. The other volumes, *The Two Towers* and *The Return of the King*, were published in 1955. Tolkien was later to revise the whole work for a hardback second edition (1966).

back to the *Odyssey* and beyond—it makes not a return but an advance or revolution: the conquest of new territory.

Nothing quite like it was ever done before. 'One takes it', says Naomi Mitchison, 'as seriously as Malory'.[2] But then the ineluctable sense of reality which we feel in the *Morte d'Arthur* comes largely from the great weight of other men's work built up century by century, which has gone into it. The utterly new achievement of Professor Tolkien is that he carries a comparable sense of reality unaided. Probably no book yet written in the world is quite such a radical instance of what its author has ere called 'sub-creation'.[3] The direct debt (there are of course subtler kinds of debt) which every author must owe to the actual universe is here deliberately

2 'One Ring to Bind Them', *New Statesman and Nation* (18 September 1954).

3 'On Fairy-Stories' in *Essays Presented to Charles Williams* (1947).

reduced to the minimum. Not content to create his own story, he creates, with an almost insolent prodigality, the whole world in which it is to move, with its own theology, myths, geography, history, paleography, languages, and orders of beings—a world 'full of strange creatures beyond count'.[4] The names alone are a feast, whether redolent of quiet countryside (Michel Delving, South Farthing), tall and kingly (Boromir, Faramir, Elendil), loathsome like Smeagol, who is also Gollum, or frowning in the evil strength of Barad Dur or Gorgoroth; yet best of all (Lothlórien, Gilthoniel, Galadriel) when they embody that piercing, high elvish beauty of which no other prose writer has captured so much.

Such a book has of course its predestined readers, even now more numerous and more critical than is always realised. To them a reviewer need say little, except that here are beauties which pierce like swords

4 'Prologue', *The Fellowship of the Ring*.

or burn like cold iron; here is a book that will break your heart. They will know that this is good news, good beyond hope. To complete their happiness one need only add that it promises to be gloriously long: this volume is only the first of three. But it is too great a book to rule only its natural subjects. Something must be said to 'those without', to the unconverted. At the very least, possible misunderstandings may be got out of the way.

First, we must clearly understand that though *The Fellowship* in one way continues its author's fairy tale, *The Hobbit*, it is in no sense an overgrown 'juvenile'. The truth is the other way round. *The Hobbit* was merely a fragment torn from the author's huge myth and adapted for children; inevitably losing something by the adaptation. *The Fellowship* gives us at last the lineaments of that myth 'in their true dimensions like themselves'. Misunderstanding on this point might easily be encouraged by the first chapter, in which the author (taking a risk) writes almost in the manner of the earlier and far lighter book. With some who will find the main body of the book deeply moving, this chapter may not be a favourite.

Yet there were good reasons for such an opening; still more for the Prologue (wholly admirable, this) which precedes it. It is essential that we should first be well steeped in the 'homeliness', the frivolity, even (in its best sense) the vulgarity of the creatures called Hobbits; these unambitious folk, peaceable yet almost anarchical, with faces 'good-natured rather than beautiful' and 'mouths apt to laughter and eating',[5] who treat smoking as an art and like books which tell them what they already know. They are not an allegory of the English, but they are perhaps a myth that only an Englishman (or, should we add, a Dutchman?) could have created. Almost the central theme of the book is the contrast between the

5 'Prologue', *The Fellowship of the Ring.*

Hobbits (or 'the Shire') and the appalling destiny to which some of them are called, the terrifying discovery that the humdrum happiness of the Shire, which they had taken for granted as something normal, is in reality a sort of local and temporary accident, that its existence depends on being protected by powers which Hobbits dare not imagine, that any Hobbit may find himself forced out of the Shire and caught up into that high conflict. More strangely still, the event of that conflict between strongest things may come to depend on him, who is almost the weakest.

What shows that we are reading myth, not allegory, is that there are no pointers to a specifically theological, or political, or psychological application. A myth points, for each reader, to the realm he lives in most. It is a master key; use it on what door you like. And there are other themes in *The Fellowship* equally serious.

That is why no catchwords about 'escapism' or 'nostalgia' and no distrust of 'private worlds' are in court. This is no Angria, no dreaming; it is sane and vigilant invention, revealing at point after point the integration of the author's mind. What is the use of calling 'private' a world we can all walk into and test and in which we find such a balance? As for escapism, what we chiefly escape is the illusions of our ordinary life. We certainly do not escape anguish. Despite many a snug fireside and many an hour of good cheer to gratify the Hobbit in each of us, anguish is, for me, almost the prevailing note. But not, as in the literature most typical of our age, the anguish of abnormal or contorted souls: rather that anguish of those who were happy before a certain darkness came up and will be happy if they live to see it gone.

Nostalgia does indeed come in; not ours nor the author's, but that of the characters. . . . Our own

world, except at certain rare moments, hardly seems so heavy with its past. This is one element in the anguish which the characters bear. But with the anguish there comes also a strange exaltation. They are at once stricken and upheld by the memory of vanished civilizations and lost splendour. They have outlived the second and third Ages; the wine of life was drawn long since. As we read we find ourselves sharing their burden; when we have finished, we return to our own life not relaxed but fortified.

But there is more in the book still. Every now and then, risen from sources we can only conjecture and almost alien (one would think) to the author's habitual imagination, figures meet us so brimming with life (not human life) that they make our sort of anguish and our sort of exaltation seem unimportant. Such is Tom Bombadil, such the unforgettable Ents. This is surely the utmost reach of invention, when an author produces what seems to be not even his own, much less anyone else's. Is mythopoeia, after all, not the most, but the least, subjective of activities?

Even now I have left out almost everything—the silvan leafiness, the passions, the high virtues, the remote horizons. Even if I had space I could hardly convey them. And after all the most obvious appeal of the book is perhaps also its deepest: 'there was sorrow then too, and gathering dark, but great valor, and great deeds that were not wholly vain'.[6] *Not wholly vain*—it is the cool middle point between illusion and disillusionment.

When I reviewed the first volume of this work I hardly dared to hope it would have the success which I was sure it deserved. Happily I am proved wrong. There is, however, one piece of false criticism which had better be answered; the complaint that the characters are all either black or white. Since the climax

---

6 *The Fellowship of the Ring*, Bk. I, ch. 2.

of Volume I was mainly concerned with the struggle between good and evil in the mind of Boromir, it is not easy to see how anyone could have said this. I will hazard a guess. 'How shall a man judge what to do in such times?' asks someone in Volume II. 'As he has ever judged', comes the reply. 'Good and ill have not changed . . . nor are they one thing among Elves and Dwarves and another among Men.'[7]

This is the basis of the whole Tolkienian world. I think some readers, seeing (and disliking) this rigid demarcation of black and white, imagine they have seen a rigid demarcation between black and white people. Looking at the squares, they assume (in defiance of the *facts*) that all the pieces must be making bishops' moves which confine them to one colour. But even such readers will hardly brazen it out through the two last volumes. Motives, even in the right side, are mixed. Those who are now traitors

7 *The Two Towers*, Bk. III, ch. 2.

usually began with comparatively innocent intentions. Heroic Rohan and imperial Gondor are partly diseased. Even the wretched Smeagol, till quite late in the story, has good impulses; and (by a tragic paradox) what finally pushes him over the brink is an unpremeditated speech by the most selfless character of all. . . .

Of picking out great moments (such as the cock-crow at the Siege of Gondor) there would be no end; I will mention two general (and totally different) excellences. One, surprisingly, is realisms. This war has the very quality of the war my generation knew. It is all here: the endless, unintelligible movement, the sinister quiet of the front when 'everything is now ready', the flying civilians, the lively, vivid friendships, the background of something like despair and the merry foreground, and such heaven-sent windfalls as a *cache* of choice tobacco 'salvaged' from a ruin. The author has told us elsewhere that his taste for fairy tale was wakened into maturity by active service;[8] that, no doubt, is why we can say of his war scenes (quoting Gimli the Dwarf), 'There is good rock here. This country has tough bones'.[9] The other excellence is that no individual, and no species, seems to exist only for the sake of the plot. All exist in their own right and would have been worth creating for their mere flavour even if they had been irrelevant. Treebeard would have served any other author (if any other could have conceived him) for a whole book. His eyes are 'filled up with ages of memory and long, slow, steady thinking'.[10] Through those ages his name has grown with him, so that he cannot now tell it; it would, by now, take too long to pronounce. When he learns that the thing they are standing on

8 'On Fairy-Stories'.

9 *The Two Towers*, Bk. III, ch. 2.

10 *The Two Towers*, Bk. III, ch. 4.

is a hill, he complains that this is but 'a hasty word'[11] for that which has so much history in it.

How far Treebeard can be regarded as a 'portrait of the artist' must remain doubtful; but when he hears that some people want to identify the Ring with the hydrogen bomb, and Mordor with Russia, I think he might call it a 'hasty' word. How long do people think a world like his takes to grow? Do they think it can be done as quickly as a modern nation changes its Public Enemy Number One or as modern scientists invent new weapons? When Professor Tolkien began there was probably no nuclear fission and the contemporary incarnation of Mordor was a good deal nearer our shores. But the text itself teaches us that Sauron is eternal; the war of the Ring is only one of a thousand wars against him. Every time we shall be wise to fear his ultimate victory, after which there will be 'no more songs'. Again and

---

11 *The Two Towers*, Bk. III, ch. 4.

again we shall have good evidence that 'the wind is setting East, and the withering of all woods may be drawing near'.[12] Every time we win we shall know that our victory is impermanent. If we insist on asking for the moral of the story, that is its moral: a recall from facile optimism and wailing pessimism alike, to that hard, yet not quite desperate, insight into Man's unchanging predicament by which heroic ages have lived. It is here that the Norse affinity is strongest; hammer-strokes, but with compassion.

'But why', (some ask), 'why, if you have a serious comment to make on the real life of men, must you do it by talking about a phantasmagoric never-never land of your own?' Because, I take it, one of the main things the author wants to say is that the real life of men is of that mythical and heroic quality. One can see the principle at work in his characterization. Much that in a realistic work would be done

---

12 *The Two Towers*, Bk. III, ch. 4.

by 'character delineation' is here done simply by making the character an elf, a dwarf, or a hobbit. The imagined beings have their insides on the outside; they are visible souls. And Man as a whole, Man pitted against the universe, have we seen him at all till we see that he is like a hero in a fairy tale? In the book Eomer rashly contrasts 'the green earth' with 'legends'. Aragorn replies that the green earth itself is 'a mighty matter of legend'.[13]

The value of the myth is that it takes all the things we know and restores to them the rich significance which has been hidden by 'the veil of familiarity'. The child enjoys his cold meat (otherwise dull to him) by pretending it is buffalo, just killed with his own bow and arrow. And the child is wise. The real meat comes back to him more savoury for having been dipped in a story; you might say that only then is it the real meat. If you are tired of the real land-

---

13 *The Two Towers*, Bk. III, ch. 2.

scape, look at it in a mirror. By putting bread, gold, horse, apple, or the very roads into a myth, we do not retreat from reality: we rediscover it. As long as the story lingers in our mind, the real things are more themselves. This book applies the treatment not only to bread or apple but to good and evil, to our endless perils, our anguish, and our joys. By dipping them in myth we see them more clearly. I do not think he could have done it in any other way.

The book is too original and too opulent for any final judgment on a first reading. But we know at once that it has done things to us. We are not quite the same men. And though we must ration ourselves in our re-readings, I have little doubt that the book will soon take its place among the indispensables.

Dear Tollers,

I have been trying—like a boy with a bit of toffee—to take Vol. I slowly, to make it last, but appetite overmastered me and it's now finished: far too short for me. The spell does not break. The love of Gimli and the departure from Lothlórien is still almost unbearable. What came out stronger at this reading than on any previous one was the gradual coming of the shadow—step by step—over Boromir.

—Letter to J. R. R. Tolkien, December 7, 1953

ON THE DANGERS OF CONFUSING SAGA WITH HISTORY

THE ACTUAL HISTORY OF EVERY COUNTRY IS FULL OF shabby and even shameful doings. The heroic stories, if taken to be typical, give a false impression of it and are often themselves open to serious historical criticism. Hence a patriotism based on our glorious past is fair game for the debunker. As knowledge increases it may snap and be converted into disillusioned cynicism, or may be maintained by a voluntary shutting of the eyes. But who can condemn what clearly makes many people, at many important

---

The Four Loves

(From Chapter II, "Likings and Loves for the Sub-human")

moments, behave so much better than they could have done without its help?

I think it is possible to be strengthened by the image of the past without being either deceived or puffed up. The image becomes dangerous in the precise degree to which it is mistaken, or substituted, for serious and systematic historical study. The stories are best when they are handed on and accepted as stories. I do not mean by this that they should be handed on as mere fictions (some of them are after all true). But the emphasis should be on the tale as such, on the picture which fires the imagination, the example that strengthens the will. The schoolboy who hears them should dimly feel—though of course he cannot put it into words—that he is hearing *saga*. Let him be thrilled—preferably 'out of school'—by the 'Deeds that won the Empire'; but the less we mix this up with his 'history lessons' or mistake it for a serious analysis—worse still, a justification—of

What does seem to me poisonous, what breeds a type of patriotism that is pernicious if it lasts but not likely to last long in an educated adult, is the perfectly serious indoctrination of the young in knowably false or biased history—the heroic legend drably disguised as text-book fact.

imperial policy, the better. When I was a child I had a book full of coloured pictures called *Our Island Story*. That title has always seemed to me to strike exactly the right note. The book did not look at all like a text-book either. What does seem to me poisonous, what breeds a type of patriotism that is pernicious if it lasts but not likely to last long in an educated adult, is the perfectly serious indoctrination of the young in knowably false or biased history—the heroic legend drably disguised as text-book fact. With this creeps in the tacit assumption that other nations have not equally their heroes; perhaps even the belief—surely it is very bad biology—that we can literally 'inherit' a tradition. And these almost inevitably lead on to a third thing that is sometimes called patriotism.

This third thing is not a sentiment but a belief: a firm, even prosaic belief that our own nation, in sober fact, has long been, and still is markedly superior to all others. I once ventured to say to an old clergyman who was voicing this sort of patriotism, 'But, sir, aren't we told that *every* people thinks its own men the bravest and its own women the fairest in the world?' He replied with total gravity—he could not have been graver if he had been saying the Creed at the altar—'Yes, but in England it's true.' To be sure, this conviction had not made my friend (God rest his soul) a villain; only an extremely lovable old ass. It can however produce asses that kick and bite. On the lunatic fringe it may shade off into that popular Racialism which Christianity and science equally forbid.

ON TWO WAYS OF TRAVELING AND TWO WAYS OF READING

THERE ARE TWO WAYS OF ENJOYING THE PAST, AS there are two ways of enjoying a foreign country. One man carries his Englishry abroad with him and brings it home unchanged. Wherever he goes he consorts with other English tourists. By a good hotel he means one that is like an English hotel. He complains of the bad tea where he might have had excellent coffee. . . .

But there is another sort of travelling and another sort of reading. You can eat the local food and drink

---

Studies in Medieval and Renaissance Literature
(from "De Audiendis Poetis")

the local wines, you can share the foreign life, you can begin to see the foreign country as it looks, not to the tourist, but to its inhabitants. You can come

It would seem to me a waste of the past if we were content to see in the literature of every bygone age only the reflection of our own faces.

home modified, thinking and feeling as you did not think and feel before. So with the old literature. You can go beyond the first impression that a poem makes on your modern sensibility. By study of things outside the poem, by comparing it with other poems, by steeping yourself in the vanished period, you can

then re-enter the poem with eyes more like those of the natives; now perhaps seeing that the associations you gave to the old words were false, that the real implications were different than you supposed, that what you thought strange was then ordinary and that what seemed to you ordinary was then strange. . . .

I am writing to help, if I may, the second sort of reading. Partly, of course, because I have a historical motive. I am a man as well as a lover of poetry: being human, I am inquisitive, I want to know as well as to enjoy. But even if enjoyment alone were my aim I should still choose this way, for I should hope to be led by it to newer and fresher enjoyments, things I could never have met in my own period, modes of feeling, flavours, journey into the real past. I have lived nearly sixty years with myself and my own century and am not so enamored of either as to desire no glimpse of a world beyond them. As the mere tourist's kind of holiday abroad seems to me rather a waste of Europe—there is more to be got out of it than he gets—so it would seem to me a waste of the past if we were content to see in the literature of every bygone age only the reflection of our own faces.

Part Two

SHORT READINGS ON READING

WORD COMBINATIONS

Isn't it funny the way some combinations of words can give you—almost apart from their meaning—a thrill like music? It is because I know that you can feel this magic of words AS words that I do not despair of teaching you to appreciate poetry: or rather to appreciate all good poetry, as you now appreciate some.

Letter to Arthur Greeves, March 21, 1916

SINCERITY AND TALENT

We must not say that Bunyan wrote well because he was a sincere, forthright man who had no literary affectations and simply said what he meant. I do not doubt that is the account of the matter that Bunyan would have given himself. But it will not do. If it were the real explanation, then every sincere, forthright, unaffected man could write as well. But most people of my age learned from censoring the letters of the troops, when we were subalterns in the first war, that unliterary people, however sincere and forthright in their talk, no sooner take a pen in hand than cliché and platitude flow from it. The shocking truth is that, while insincerity may be fatal to good writing, sincerity, of itself, never taught anyone to write well. It is a moral virtue, not a literary talent. We may hope it is rewarded in a better world: it is not rewarded on Parnassus.

Selected Literary Essays
(from "The Vision of John Bunyan")

PROSE STYLE

You have started the question of prose style in your letter and ask whether it is anything more than the 'literal meaning of the words'. On the contrary it means less—it means the words themselves. For every thought can be expressed in a number of different ways: and style is the art of expressing a given thought in the most beautiful words and rhythms of words. For instance a man might say 'When the constellations which appear at early morning joined in musical exercises and the angelic spirits loudly testified to their satisfaction'. Expressing exactly the same thought, the Authorized Version says 'When the morning stars sang together and all the sons of God shouted for joy'. Thus by the power of style what was nonsense becomes ineffably beautiful.

Letter to Arthur Greeves, August 4, 1917

NOT *IN* BUT *THROUGH*

The books or the music in which we thought the beauty was located will betray us if we trust to them; it was not *in* them, it only came *through* them, and what came through them was longing.

The Weight of Glory and Other Addresses
(from "The Weight of Glory")

PLEASURE

For a great deal (not all) of our literature was made to be read lightly, for entertainment. If we do not read it, in a sense, 'for fun' and with our feet on the fender, we are not using it as it was meant to be used, and all our criticism of it will be pure illusion. For you cannot judge any artifact except by using it as it was intended. It is no good judging a butter-knife by seeing whether it will saw logs. Much bad criticism, indeed, results from the efforts of critics to get a work-time result out of something that never aimed at producing more than pleasure.

Christian Reflections

(from "Christianity and Culture")

ORIGINALITY

No man who values originality will ever be original. But try to tell the truth as you see it, try to do any bit of work as well as it can be done for the work's sake, and what men call originality will come unsought.

The Weight of Glory
(from "Membership")

THE UP-TO-DATE MYTH

The more up to date the Book is, the sooner it will be dated.

Letters to Malcolm: Chiefly on Prayer
(from Chapter 2)

KEEPING UP

Incidentally, what is the point of keeping in touch with the contemporary scene? Why should one read authors one doesn't like because they happen to be alive at the same time as oneself? One might as well read everyone who had the same job or the same coloured hair, or the same income, or the same chest measurements, as far as I can see.

Letter to Ruth Pitter, January 6, 1951

WIDE TASTES

By having a great many friends I do not prove that I have a wide appreciation of human excellence. You might as well say I prove the width of my literary taste by being able to enjoy all the books in my own study. The answer is the same in both cases—'You chose those books. You chose those friends. Of course they suit you.' The truly wide taste in reading is that which enables a man to find something for his needs on the sixpenny tray outside any secondhand bookshop. The truly wide taste in humanity will similarly find something to appreciate in the cross-section of humanity whom one has to meet every day.

The Four Loves

(from Chapter III, "Affections")

REAL ENJOYMENT

After a certain kind of sherry party, where there have been cataracts of culture but never one word or one glance that suggested a real enjoyment of any art, any person, or any natural object, my heart warms to the schoolboy on the bus who is reading Fantasy and Science Fiction, rapt and oblivious of all the world beside. For here also I should feel that I had met something real and live and unfabricated; genuine literary experience, spontaneous and compulsive, disinterested. I should have hopes of that boy. Those who have greatly cared for any book whatever may possibly come to care, some day, for good books. The organs of appreciation exist in them. They are not impotent. And even if this particular boy is never going to like anything severer than science-fiction, even so,

> The child whose love is here, at least doth reap
> One precious gain, that he forgets himself.

The World's Last Night
(from "The Lilies That Fester")

LITERARY SNOBS

Some critics write of those who constitute the literary 'many' as if they belonged to the many in every respect, and indeed to the rabble. They accuse them of illiteracy, barbarism, 'crass', 'crude' and 'stock' responses which (it is suggested) must make them clumsy and insensitive in all the relations of life and render them a permanent danger to civilisation. It sometimes sounds as if the reading of 'popular' fiction involved moral turpitude. I do not find this borne out by experience. I have a notion that these 'many' include certain people who are equal or superior to some of the few in psychological health, in moral virtue, practical prudence, good manners, and general adaptability. And we all know very well that we, the literary, include no small percentage of the ignorant, the caddish, the stunted, the warped, and the truculent. With the hasty and wholesale apartheid of those who ignore this we must have nothing to do.

An Experiment in Criticism
(from Chapter 2, "False Characteristics")

RE-READING FAVORITES EACH DECADE

Clearly one must read every good book at least once every ten years.

Letter to Arthur Greeves, August 17, 1933

READING AND EXPERIENCE

You ask me whether I have ever been in love: fool as I am, I am not quite such a fool as all that. But if one is only to talk from firsthand experience on any subject, conversation would be a very poor business. But though I have no personal experience of the thing they call love, I have what is better—the experience of Sappho, of Euripides of Catullus of Shakespeare of Spenser of Austen of Brontë of, of—anyone else I have read. We see through their eyes. And as the greater includes the less, the passion of a great mind includes all the qualities of the passion of a small one. Accordingly, we have every right to talk about it.

Letter to Arthur Greeves, October 12, 1915

FREE TO SKIP

It is a very silly idea that in reading a book you must never 'skip'. All sensible people skip freely when they come to a chapter which they find is going to be no use to them.

Mere Christianity
(from Chapter 3, "Time and Beyond Time")

FREE TO READ

The State exists simply to promote and to protect the ordinary happiness of human beings in this life. A husband and wife chatting over a fire, a couple of friends having a game of darts in a pub, a man reading a book in his own room or digging in his own garden—that is what the State is there for. And unless they are helping to increase and prolong and protect such moments, all the laws, parliaments, armies, courts, police, economics, etc., are simply a waste of time.

Mere Christianity
(from Chapter 8, "Is Christianity Hard or Easy?")

HUCK

I have been regaling myself on Tom Sawyer and Huckleberry Finn. I wonder why that man never wrote anything else on the same level? The scene in which Huck decides to be 'good' by betraying Jim, and then finds he can't and concludes that he is a reprobate, is really unparalleled in humour, pathos, & tenderness. And it goes down to the very depth of all moral problems.

Letter to Warfield M. Firor (BOD), December 6, 1950

THE GLORIES OF CHILDHOOD—VERSUS ADOLESCENCE

About re-reading books: I find like you that those read in my earlier 'teens often have no appeal, but this is not nearly so often true of those read in earlier childhood. Girls may develop differently, but for me, looking back, it seems that the glories of childhood and the glories of adolescence are separated by a howling desert during which one was simply a greedy, cruel, spiteful little animal and imagination, in all but the lowest form, was asleep.

Letter to Rhona Bodle, December 26, 1953

JANE AUSTEN

I am glad you think J. Austen a sound moralist. I agree. And not platitudinous, but subtle as well as firm.

Letter to Dom Bede Griffiths, May 5, 1952

I don't believe anything will keep the right reader & the right book apart. But our literary loves are as diverse as our human! You couldn't make me like Henry James or dislike Jane Austen whatever you did.

Letter to Rhona Bodle, September 14, 1953

I've been reading *Pride and Prejudice* on and off all my life and it doesn't wear out a bit.

Letter to Sarah Neylan, January 16, 1954

ART AND LITERATURE

I do most thoroughly agree with what you say about Art and Literature. To my mind they can only be healthy when they are either (a) admittedly aiming at nothing but innocent recreation or (b) definitely the handmaids of religious or at least moral truth. Dante is alright and Pickwick is alright. But the great *serious irreligious* art—art for art's sake—is all balderdash; and incidentally never exists when art is really flourishing. One can say of Arts as an author I recently read said of love (sexual love I mean), 'It ceases to be a devil when it ceases to be god.' Isn't that well put?

Letter to Dom Bede Griffiths, April 16, 1940

ART APPRECIATION

Many modern novels, poems, and pictures, which we are brow-beaten into 'appreciating', are not good work because they are not *work* at all. They are mere puddles of spilled sensibility or reflection. When an artist is in the strict sense working, he of course takes into account the existing taste, interests, and capacity of his audience. These, no less than the language, the marble, or the paint, are part of his raw materials, to be used, tamed, sublimated, not ignored or defied. Haughty indifference to them is not genius or integrity; it is laziness and incompetence.

The World's Last Night and Other Essays
(from "Good Work and Good Works")

LOOK. LISTEN. RECEIVE.

The first demand any work of any art makes upon us is surrender. Look. Listen. Receive. Get yourself out of the way. (There is no good asking first whether the work before you deserves such a surrender, for until you have surrendered you cannot possibly find out.)

An Experiment in Criticism

(from Chapter 3, "How the Few and the Many Use Pictures and Music")

TALKING ABOUT BOOKS

When one has read a book, I think there is nothing so nice as discussing it with some one else—even though it sometimes produces rather fierce arguments.

Letter to Arthur Greeves, March 14, 1916

THE BLESSING OF CORRESPONDENCE

It is the immemorial privilege of letter-writers to commit to paper things they would not say: to write in a more grandiose manner than that in which they speak: and to enlarge upon feelings which would be passed by unnoticed in conversation.

Letter to Arthur Greeves, November 10, 1914

IN PRAISE OF DANTE

I think Dante's poetry, on the whole, the greatest of all the poetry I have read: yet when it is at its highest pitch of excellence, I hardly feel that Dante has very much to do. There is a curious feeling that the great poem is writing itself, or at most, that the tiny figure of the poet is merely giving the gentlest guiding touch, here and there, to energies which, for the most part, spontaneously group themselves and perform the delicate evolutions which make up the Comedy. . . . I draw the conclusion that the highest reach of the whole poetic art turns out to be a kind of abdication, and is attained when the whole image of the world the poet sees has entered so deeply into his mind that henceforth he has only to get himself out of the way, to let the seas roll and the mountains shake their leaves or the light shine and the spheres revolve, and all this will *be* poetry, not things you write poetry about.

Studies in Medieval and Renaissance Literature
(from "Dante's Similes")

ON ALEXANDRE DUMAS

I tried, at W's [Lewis's brother Warren's] earnest recommendation, to read the *Three Musketeers,* but not only got tired but also found it disgusting. All of these swaggering bullies, living on the money of their mistresses—faugh! . . . You are in an abstract world of gallantry and adventure which has no *roots*—no connection with human nature or mother earth. When the scene shifts from Paris to London there is no sense that you have reached a new country, no change of atmosphere. I don't think there is a single passage to show that Dumas had ever seen a cloud, a road, or a tree.

Letter to Arthur Greeves, March 25, 1933

THE DELIGHT OF FAIRY TALES

Curiously enough it is at this time [age 12], not in earlier childhood, that I chiefly remember delighting in fairy tales. I fell deeply under the spell of Dwarfs—the old bright-hooded, snowy-bearded dwarfs we had in those days before Arthur Rackham sublimed, or Walt Disney vulgarized, the earthmen. I visualized them so intensely that I came to the very frontiers of hallucination; once, walking in the garden, I was for a second not quite sure that a little man had not run past me into the shrubbery. I was faintly alarmed, but it was not like my night fears. A fear that guarded the road to Faerie was one I could face. No one is a coward at all points.

Surprised by Joy

(from Chapter III, "Mountbracken and Campbell")

LANGUAGE AS COMMENT

Mere description is impossible. Language forces you to an implicit comment.

Present Concerns
(from "Prudery and Philology")

COMMUNICATING THE ESSENCE OF OUR LIVES

The very essence of our life as conscious beings, all day and every day, consists of something which cannot be communicated except by hints, similes, metaphors, and the use of those emotions (themselves not very important) which are pointers to it.

Christian Reflections

(from "The Language of Religion")

MAPPING MY BOOKS

To enjoy a book like that thoroughly I find I have to treat it as a sort of hobby and set about it seriously. I begin by making a map on one of the end leafs: then I put in a genealogical tree or two. Then I put a running headline at the top of each page: finally I index at the end all the passages I have for any reason underlined. I often wonder—considering how people enjoy themselves developing photos or making scrapbooks—why so few people make a hobby of their reading in this way. Many an otherwise dull book which I had to read have I enjoyed in this way, with a fine-nibbed pen in my hand: one is *making* something all the time and book so read acquires the charm of a toy without losing that of a book.

Letter to Arthur Greeves, February 1932

ON PLATO AND ARISTOTLE

To lose what I owe to Plato and Aristotle would be like an amputation of a limb.

Rehabilitations and Other Essays
(from "The Idea of an 'English School' ")

IMAGINATION

It seems to me appropriate, almost inevitable, that when that great Imagination which in the beginning, for Its own delight and for the delight of men and angels and (in their proper mode) of beasts, had invented and formed the whole world of Nature, submitted to express Itself in human speech, that speech should sometimes be poetry. For poetry too is a little incarnation, giving body to what had been before invisible and inaudible.

Reflections on the Psalms
(from Chapter I, "Introductory")

IF ONLY

If only one had time to read a little more: we either get shallow & broad or narrow and deep.

Letter to Arthur Greeves, March 2, 1919

ON SHAKESPEARE

Where Milton marches steadily forward, Shakespeare behaves rather like a swallow. He darts at the subject and glances away; and then he is back again before your eyes can follow him. It is as if he kept on having tries at it, and being dissatisfied. He darts image after image at you and still seems to think that he has not done enough. He brings up a whole light artillery of mythology, and gets tired of each piece almost before he has fired it. He wants to see the object from a dozen different angles; if the undignified word is pardonable, he *nibbles*, like a man trying a tough biscuit now from this side and now from that. You can find the same sort of contrast almost anywhere between these two poets.

Selected Literary Essays

(from "Variation in Shakespeare and Others")

ON *HAMLET*

'Most certainly an artistic failure.' All argument is for that conclusion—until you read or see *Hamlet* again. And when you do, you are left saying that if this is failure, then failure is better than success. We want more of these 'bad' plays. From our first childish reading of the ghost scenes down to those golden minutes which we stole from marking examination papers on *Hamlet* to read a few pages of *Hamlet* itself, have we ever known the day or the hour when its enchantment failed? . . . It has a taste of its own, an all-pervading relish which we recognize even in its smallest fragments, and which, once tasted, we recur to. When we want that taste, no other book will do instead.

Selected Literary Essays
(from "Hamlet: The Prince or The Poem?")

ON LEO TOLSTOY

The most interesting thing that has happened to me since I last wrote is reading *War and Peace*. . . . It has completely changed my view of novels.

Hitherto I had always looked on them as rather a *dangerous* form—I mean dangerous to the health of literature on the whole. I thought that the strong 'narrative lust'—the passionate itch to 'see what happened in the end'—which novels aroused, necessarily injured the taste for other, better, but less irresistible, forms of literary pleasure: and that growth of novel reading largely explained the deplorable division of readers into low-brow and high-brow—the low-brow being simply those who had learned to expect from books this 'narrative lust,' from the time they began to read, and who had thus destroyed in advance their possible taste for better things. . . .

Tolstoy, in this book, has changed all that.

Letter to Arthur Greeves, March 29, 1931

ADVICE FOR WRITING

The way for a person to develop a style is (a) to know exactly what he wants to say, and (b) to be sure he is saying exactly that. The reader, we must remember, does not start by knowing what we mean. If our words are ambiguous, our meaning will escape him. I sometimes think that writing is like driving sheep down a road. If there is any gate open to the left or the right the reader will most certainly go into it.

God in the Dock
(from "Cross-Examination")

GOOD READING

A good shoe is a shoe you don't notice. Good reading becomes possible when you need not consciously think about eyes, or light, or print, or spelling.

Letters to Malcolm: Chiefly on Prayer
(from Chapter 1)

Appendix

JOURNAL EXERCISES FOR REFLECTING ON YOUR READING LIFE

- List the ten books that have most shaped who you are today and write down a few sentences per book of how they have shaped you.

- Lewis often describes the gift of reading as the opportunity to "see through others' eyes." Which books have you read that have revealed to you a very different view of the world from your own? How did these experiences change you?

- Which books should you read that would open up other worlds you are not familiar with—allowing that the differences could be cultural, racial, religious, historical, or something else?

- Lewis highly values re-reading old books, even books from childhood. Which books have you re-read, and why did you choose to re-read them? Which books have you read more than twice? How have these books affected you?

- Write down your earliest childhood memories of books that transported you and created in you a love of books? Have you re-read these titles lately? Were they still magical? How did these early experiences influence you?

- List the "old books" you will commit to read as a break from reading all contemporary ones.

- What do you think of the genre of books called fairy stories or books of fantasy and magic—of which Lewis had much to say? Which titles have influenced you most and what do you think they have taught you about the "real" world?

- Lewis writes movingly about the discovery of his favorite author, George MacDonald. Who would you say is your favorite author, and what role has he or she played in your life?

- Lewis emphasizes the importance of reading for pleasure. Which kinds of books do you read solely for this purpose (even if they are guilty pleasures)? Why do you think this type of reading is important?

C. S. Lewis Classics

C. S. Lewis Classics

Available wherever books and e-books are sold

Discover great authors, exclusive offers, and more at hc.com.